AF228842

Narwhals live
in the sea.

5

A narwhal's home
is cold.

Superstars of the Sea
NARWHALS
Theresa Emminizer
PowerKiDS press
PK
Beginners

Let's look at narwhals!

Narwhals like to swim.

9

Narwhals have a
long tooth!

11

Narwhals have tails.

13

Narwhals are gray and white.

Narwhals come up for air.

Narwhals have families.

Narwhals swim together.

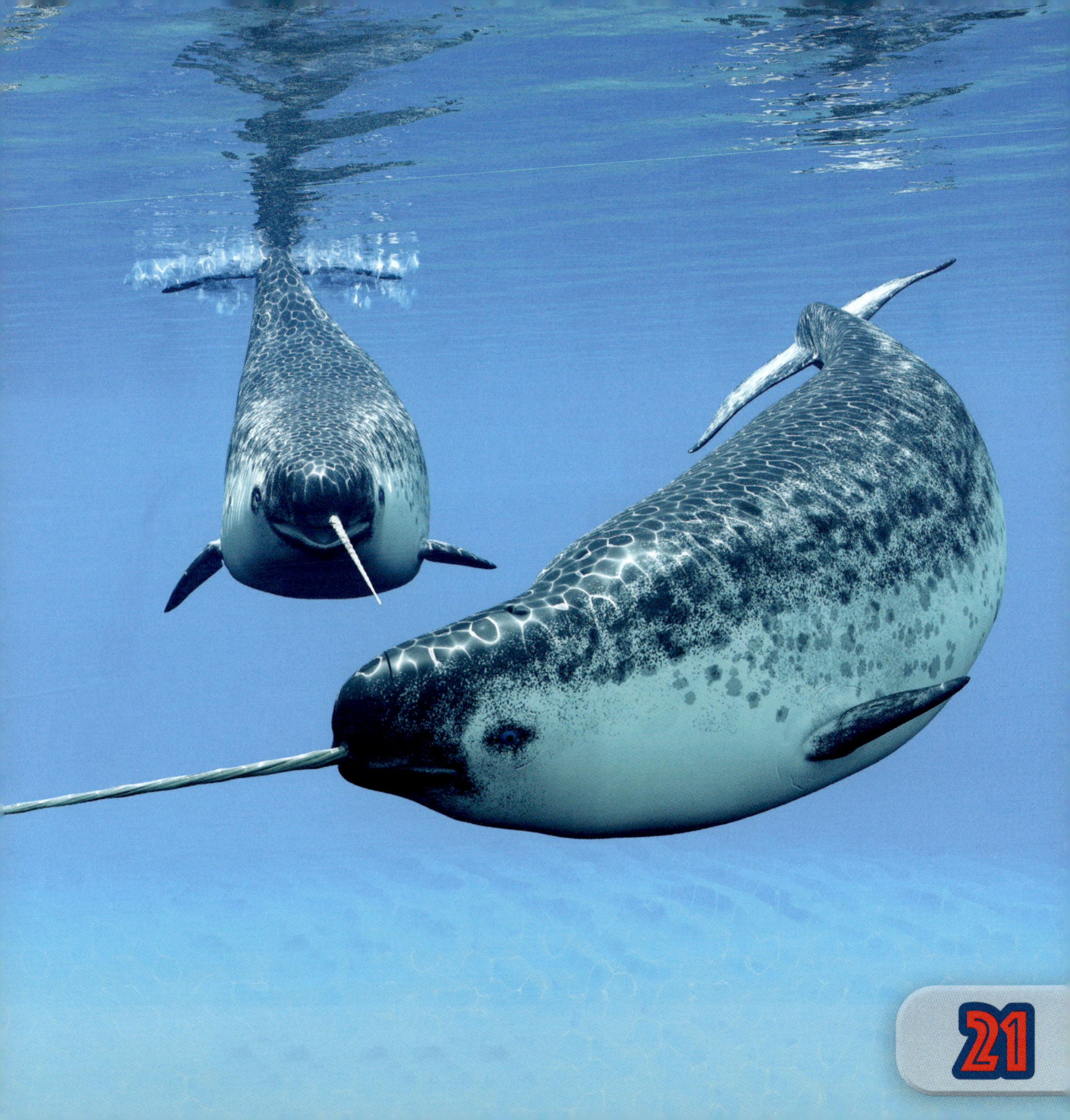

Narwhals are whales.

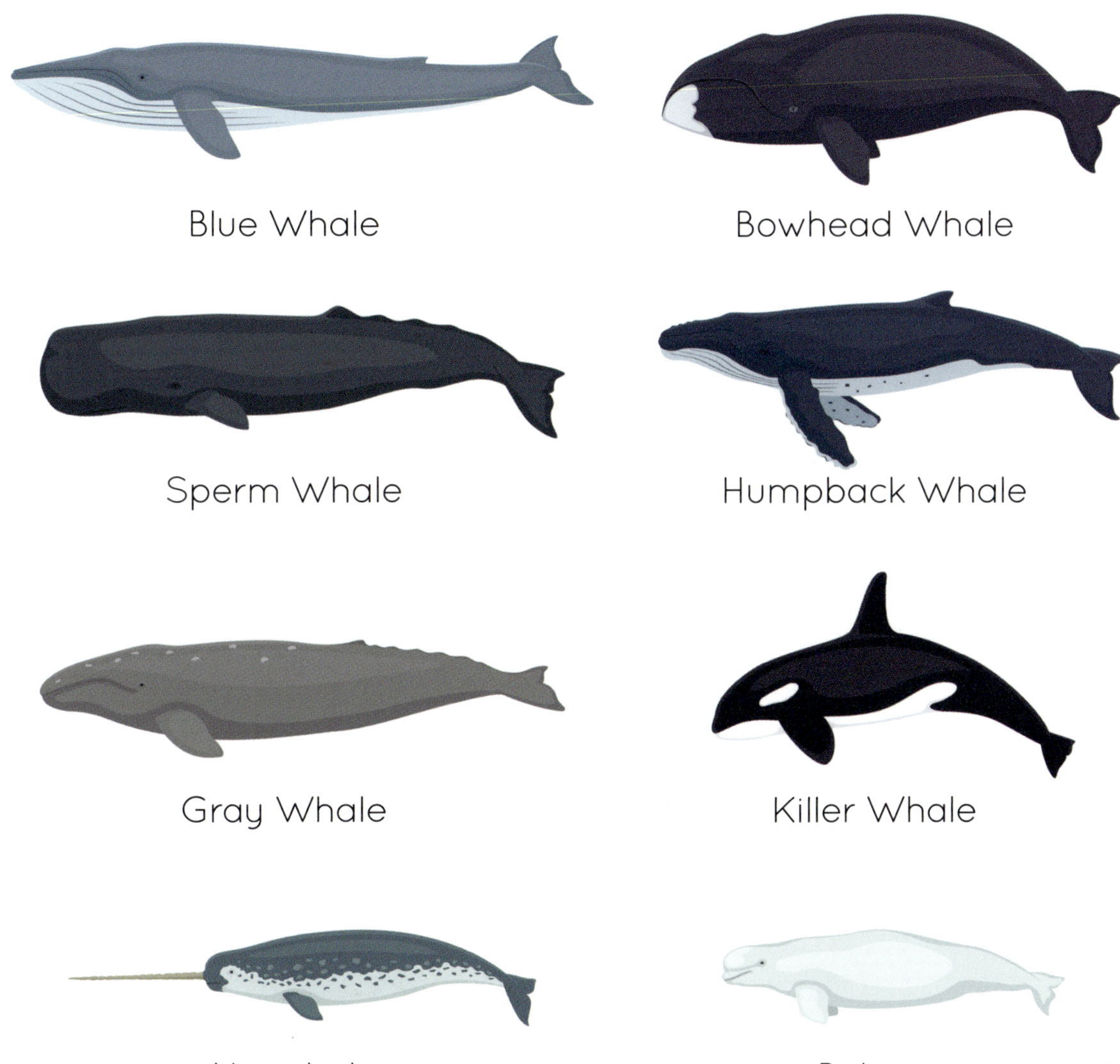

Blue Whale
Bowhead Whale
Sperm Whale
Humpback Whale
Gray Whale
Killer Whale
Narwhal
Beluga

Published in 2026 by The Rosen Publishing Group, Inc.
2544 Clinton Street, Buffalo, NY 14224

First Edition

Editor: Theresa Emminizer
Book Design: Jeffrey Taylor

Photo Credits: Cover MuhammadAsif6/Shutterstock.com; p. 3 AI generated image/Shutterstock.com; p. 5 AI generated image/Shutterstock.com; p. 7 LouieLea/Shutterstock.com; p. 9 Dotted Yeti/Shutterstock.com; p. 11 Dotted Yeti/Shutterstock.com; p. 13 Gazprom Neft/commons.wikimedia.org; p. 15 AI generated image/Shutterstock.com; p. 17 Dr. Kristin Laidre, Polar Science Center, UW NOAA/OAR/OER/commons.wikimedia.org; p. 19 Dotted Yeti/Shutterstock.com; p. 21 Catmando/Shutterstock.com; p. 23 Zhenyakot/Shutterstock.com.

Cataloging-in-Publication Data

Names: Emminizer, Theresa.
Title: Narwhals / Theresa Emminizer.
Description: Buffalo, New York : PowerKids Press, 2026. | Series: Superstars of the sea
Identifiers: ISBN 9781499451498 (pbk.) | ISBN 9781499451504 (library bound) | ISBN 9781499451511 (ebook)
Subjects: LCSH: Narwhal--Juvenile literature.
Classification: LCC QL737.C433 E46 2026 | DDC 599.5'43--dc23

Manufactured in the United States of America

CPSIA Compliance Information: Batch #CSPK26. For further information contact Rosen Publishing at 1-800-237-9932.

Find us on